Clever Critters

CAMOUFLAGE CHAMPS!

By Tracy Vonder Brink

TABLE OF CONTENTS

A Crabtree Seedlings Book

Crabtree Publishing

crabtreebooks.com

School-to-Home Support for Caregivers and Teachers

This book helps children grow by letting them practice reading. Here are a few guiding questions to help readers build comprehension skills. Possible answers appear here in red.

Before Reading:

- What do I know about this topic?
    - *I know that some animals are good at hiding.*
    - *I know that some animals use camouflage to hide from predators.*
- What do I want to learn about this topic?
    - *I want to learn which animals use camouflage.*
    - *I want to learn why camouflage is helpful.*

During Reading:

- What have I learned so far?
    - *I have learned that camouflage is an adaptation.*
    - *I have learned that adaptations help living things survive.*
- I wonder why...
    - *I wonder why a potoo sleeps during the day.*
    - *I wonder how a Caribbean reef octopus changes color.*

After Reading:

- What did I learn about this topic?
    - *I have learned that a snowshoe hare changes fur color in summer and winter.*
    - *I have learned that a decorator crab sticks rocks and seaweed to its shell.*
- Read the book again and look for the glossary words.
    - *I see the word **survive** on page 3 and the word **predators** on page 4. The other glossary words are found on pages 22 and 23.*

CAMOUFLAGE CHAMPS!

An **adaptation** is something that helps a living thing **survive**.

Blending into the background to stay safe is one adaptation. This is called **camouflage**.

These animals and plants are camouflage champs!

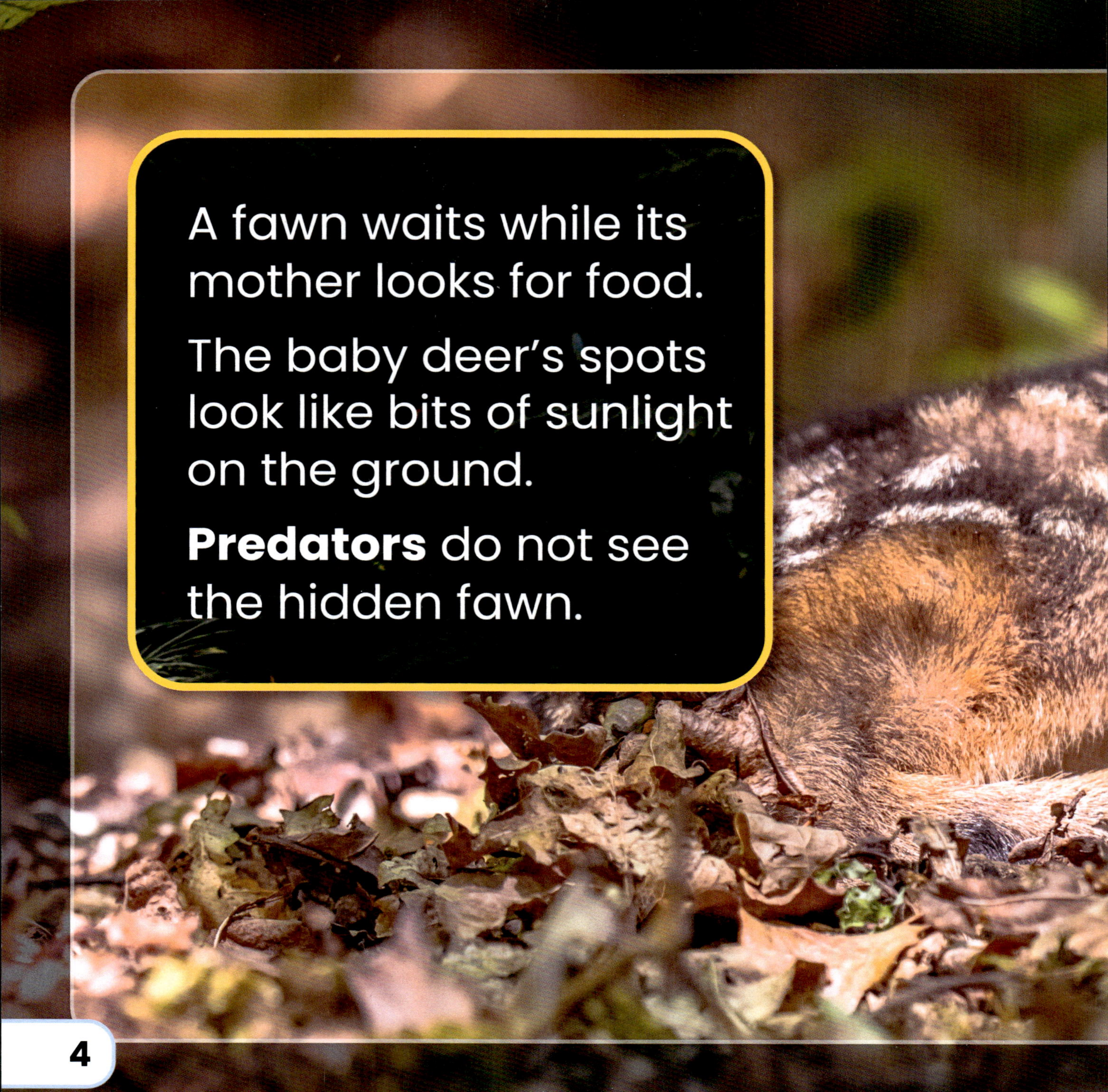

A fawn waits while its mother looks for food.

The baby deer's spots look like bits of sunlight on the ground.

Predators do not see the hidden fawn.

A walking stick is long and thin like a branch or twig.

Its color is brown like **bark**.

It holds still to hide from a hungry bird.

A potoo sleeps during the day.

Its feathers look like tree bark.

It closes its eyes and stretches its neck to look like a branch.

Potoos have unusual openings in their eyelids that let them see even when their eyes are closed.

A Caribbean reef octopus hides from a predator.

The octopus changes its skin color to match the reef.

It can even be more than one color at the same time.

A snowshoe hare changes fur color twice a year.

It is brown in summer to blend in with dirt and rocks.

It turns white in winter to match the snow.

A decorator crab picks up bits of rocks and seaweed.

It sticks the bits to **bristles** on its shell.

Its shell looks like part of the seafloor.

A gentoo penguin is black on its back and white on its belly.

Its belly looks like light shining on the water to a predator looking up.

Its back looks like dark water to a predator looking down.

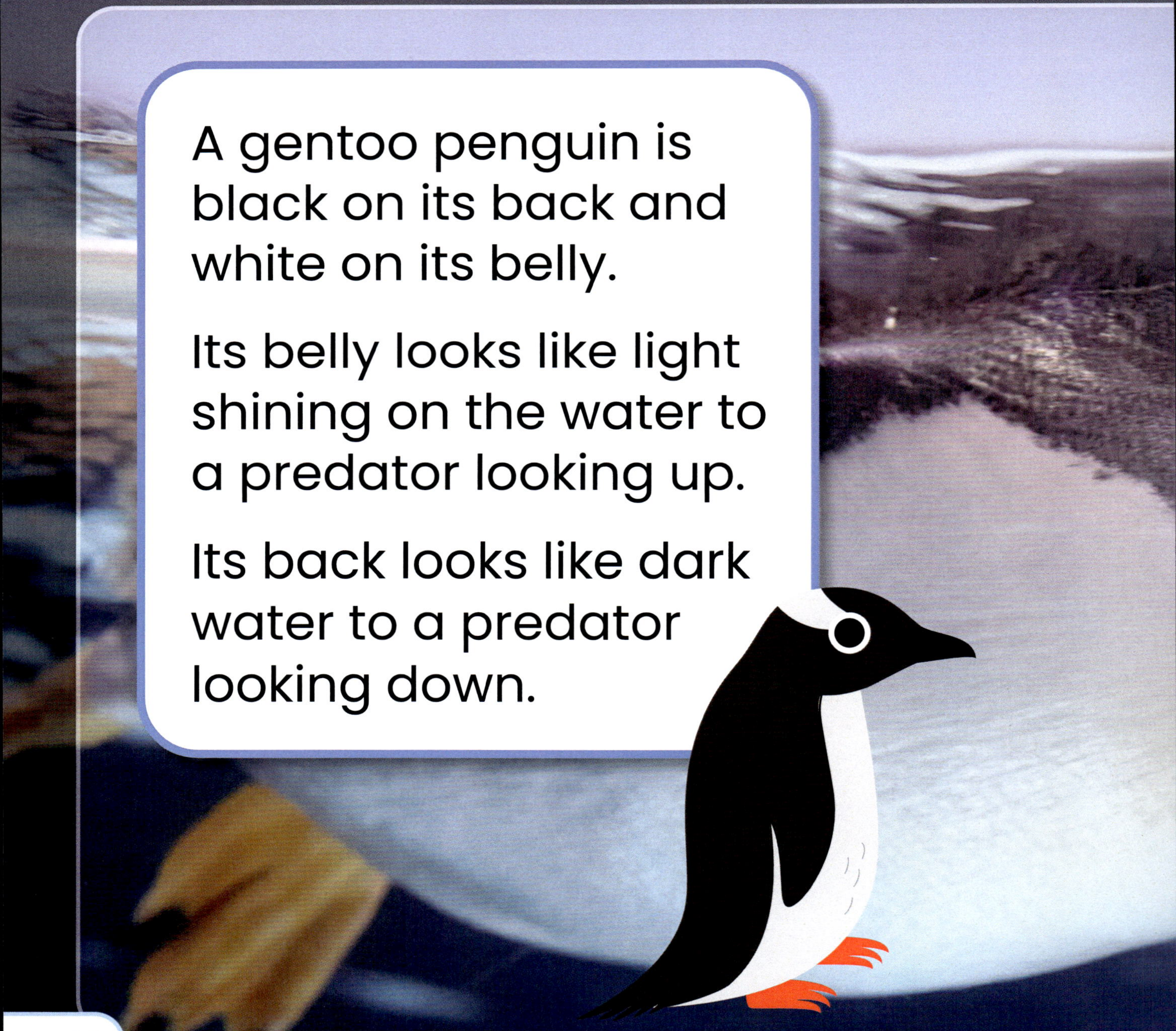

Gentoo penguins are the fastest-swimming penguins in the world!

A Lithops plant grows on the desert floor.

It looks like a rock.

A hungry animal cannot spot it.

Lithops plants are also known as living stones.

A leafy sea dragon lives on a reef where seaweed grows.

Body parts shaped like seaweed help it blend in.

Camouflage is a useful adaptation for many living things!

Glossary

adaptation (ad-uhp-TEY-shuhn): A change in an animal or a plant that helps it live in a certain place

bark (bahrk): The tough outer layer of a tree

bristles (BRIS-uhls): Short, stiff hairs that feel rough

camouflage (KAM-uh-flahzh): The way a living creature's color or shape matches its surroundings, making it hard to see

predators (PRED-uh-ters): Animals that live by hunting and eating other animals

survive (ser-VAHYV): To continue to live

Index

About the Author

Tracy Vonder Brink loves to learn about science and nature. She has written more than 100 nonfiction books for kids. Tracy lives in Cincinnati, Ohio, with her husband, two daughters, and two rescue dogs.

Written by: Tracy Vonder Brink
Designed by: Margaret Salter
Proofreader: Melissa Boyce
Photographs: All images from Shutterstock

Crabtree Publishing

crabtreebooks.com 800-387-7650

Printed in the U.S.A./CP2025

Published in Canada
Crabtree Publishing
616 Welland Avenue
St. Catharines, Ontario
L2M 5V6

Published in the United States
Crabtree Publishing
347 Fifth Avenue
Suite 1402-145
New York, New York, 10016

Library and Archives Canada Cataloguing in Publication
Available at Library and Archives Canada

Library of Congress Cataloging-in-Publication Data
Available at the Library of Congress

Hardcover: 978-1-0398-8109-9
Paperback: 978-1-0398-8470-0
Ebook (pdf): 978-1-0398-8229-4
Epub: 978-1-0398-8349-9
Read-Along: 978-1-0398-8554-7
Audio: 978-1-0398-8512-7